Address
of President Roosevelt

at the Opening of the Conference
on the Conservation of Natural Resources
at the White House
Wednesday Morning, May 13, 1908
at 10:30 o'clock

Elibron Classics
www.elibron.com

Elibron Classics series.

ISBN 0-543-70405-X (paperback)
ISBN 0-543-70404-1 (hardcover)

This Elibron Classics Replica Edition is an unabridged facsimile of the edition published in 1908 by the Government Printing Office, Washington.

Address of President Roosevelt at the opening of the Conference on the Conservation of Natural Resources, at the White House Wednesday morning, May 13 1908, at 10:30 o'clock

Washington
Government Printing Office
1908

Governors of the several States; and Gentlemen:

I welcome you to this conference at the White House. You have come hither at my request so that we may join together to consider the question of the conservation and use of the great fundamental sources of wealth of this Nation. So vital is this question, that for the first time in our history the chief executive officers of the States separately, and of the States together forming the Nation, have met to consider it.

With the governors come men from each State chosen for their special acquaintance with the terms of the problem that is before us. Among them are experts in natural resources and representatives of national organizations concerned in the development and use of these resources; the Senators and Representatives in Congress; the Supreme Court, the Cabinet, and the Inland Waterways Commission have likewise been invited to the conference, which is therefore national in a peculiar sense.

This conference on the conservation of natural resources is in effect a meeting of the representatives of all the people of the United States called to consider the weightiest problem now before the Nation; and the occasion for the meeting lies in the fact that the natural resources of our country are in danger of exhaustion if we permit the old wasteful methods of exploiting them longer to continue.

With the rise of peoples from savagery to civilization, and with the consequent growth in the extent and variety of the needs of the average man, there comes a steadily increasing growth of the amount demanded by this average man from the actual resources of the country. Yet, rather curiously, at the same time the average man is apt to lose his realization of this dependence upon nature.

Savages, and very primitive peoples generally, concern themselves only with superficial natural resources; with those which they obtain from the actual surface of the ground. As peoples become a little less primitive, their industries, although in a rude manner, are extended to resources below the surface; then, with what we call civilization and the extension of knowledge, more resources come into use, industries are multiplied, and foresight begins to become a necessary and prominent factor in life. Crops are cultivated; animals are domesticated; and metals are mastered.

Every step of the progress of mankind is marked by the discovery and use of natural resources previously unused. Without such progressive knowledge and utilization of natural resources population could not grow, nor industries multiply, nor the hidden wealth of the earth be developed for the benefit of mankind.

From the first beginnings of civilization, on the banks of the Nile and the

Euphrates, the industrial progress of the world has gone on slowly, with occasional setbacks, but on the whole steadily, through tens of centuries to the present day. But of late the rapidity of the process has increased at such a rate that more space has been actually covered during the century and a quarter occupied by our national life than during the preceding six thousand years that take us back to the earliest monuments of Egypt, to the earliest cities of the Babylonian plain.

When the founders of this nation met at Independence Hall in Philadelphia the conditions of commerce had not fundamentally changed from what they were when the Phœnician keels first furrowed the lonely waters of the Mediterranean. The differences were those of degree, not of kind, and they were not in all cases even those of degree. Mining was carried on fundamentally as it had been carried on by the Pharaohs in the countries adjacent to the Red Sea.

The wares of the merchants of Boston, of Charleston, like the wares of the merchants of Nineveh and Sidon, if they went by water, were carried by boats propelled by sails or oars; if they went by land were carried in wagons drawn by beasts of draft or in packs on the backs of beasts of burden. The ships that crossed the high seas were better than the ships that had once crossed the Ægean, but they were of the same type, after all—they were wooden ships propelled by sails; and on land, the roads were not as good as the roads of the Roman Empire, while the service of the posts was probably inferior.

In Washington's time anthracite coal was known only as a useless black stone; and the great fields of bituminous coal were undiscovered. As steam was unknown, the use of coal for power production was undreamed of. Water was practically the only source of power, save the labor of men and animals; and this power was used only in the most primitive fashion. But a few small iron deposits had been found in this country, and the use of iron by our countrymen was very small. Wood was practically the only fuel, and what lumber was sawed was consumed locally, while the forests were regarded chiefly as obstructions to settlement and cultivation.

Such was the degree of progress to which civilized mankind had attained when this nation began its career. It is almost impossible for us in this day to realize how little our Revolutionary ancestors knew of the great store of natural resources whose discovery and use have been such vital factors in the growth and greatness of this nation, and how little they required to take from this store in order to satisfy their needs.

Since then our knowledge and use of the resources of the present territory of the United States have increased a hundredfold. Indeed, the growth of this Nation by leaps and bounds makes one of the most striking and important chapters in the history of the world. Its growth has been due to the rapid development, and alas! that it should be said, to the rapid destruction, of our natural resources. Nature has supplied to us in the United States, and still supplies to us, more kinds of resources in a more lavish degree than has ever been the case at any other time or with any other people. Our position in the world has been attained by the extent and thoroughness of the control we have achieved over nature; but we are more, and not less, dependent upon what she furnishes than at any previous time of history since the days of primitive man.

Yet our fathers, though they knew so little of the resources of the country, exercised a wise forethought in reference thereto. Washington clearly saw that the perpetuity of the States could only be secured by union, and that the only feasible basis of union was an economic one; in other words, that it must be based on the development and use of their natural resources. Accordingly, he helped to outline a scheme of commercial development, and by his influence an interstate waterways commission was appointed by Virginia and Maryland.

It met near where we are now meeting, in Alexandria, adjourned to Mount Vernon, and took up the consideration of interstate commerce by the only means then available, that of water. Further conferences were arranged, first at Annapolis and then at Philadelphia. It was in Philadelphia that the representatives of all the States met for what was in its original conception merely a waterways conference; but when they had closed their deliberations the outcome was the Constitution which made the States into a Nation.

The Constitution of the United States thus grew in large part out of the necessity for united action in the wise use of one of our natural resources. The wise use of all of our natural resources, which are our national resources as well, is the great material question of to-day. I have asked you to come together now because the enormous consumption of these resources, and the threat of imminent exhaustion of some of them, due to reckless and wasteful use, once more calls for common effort, common action.

Since the days when the Constitution was adopted, steam and electricity have revolutionized the industrial world. Nowhere has the revolution been so great as in our own country. The discovery and utilization of mineral fuels and alloys have given us the lead over all other nations in the production of steel. The discovery and utilization of coal and iron have given us our railways, and have led to such industrial development as has never before been seen. The vast wealth of lumber in our forests, the riches of our soils and mines, the discovery of gold and mineral oils, combined with the efficiency of our transportation, have made the conditions of our life unparalleled in comfort and convenience.

The steadily increasing drain on these natural resources has promoted to an extraordinary degree the complexity of our industrial and social life. Moreover, this unexampled development has had a determining effect upon the character and opinions of our people. The demand for efficiency in the great task has given us vigor, effectiveness, decision, and power, and a capacity for achievement which in its own lines has never yet been matched. So great and so rapid has been our material growth that there has been a tendency to lag behind in spiritual and moral growth; but that is not the subject upon which I speak to you to-day.

Disregarding for the moment the question of moral purpose, it is safe to say that the prosperity of our people depends directly on the energy and intelligence with which our natural resources are used. It is equally clear that these resources are the final basis of national power and perpetuity. Finally, it is ominously evident that these resources are in the course of rapid exhaustion.

This Nation began with the belief that its landed possessions were illimitable and capable of supporting all the people who might care to make our country their home; but already the limit of unsettled land is in sight, and indeed but little land fitted for agriculture now remains unoccupied save what can be reclaimed by irrigation and drainage. We began with an unapproached heritage of forests; more than half of the timber is gone. We began with coal fields more extensive than those of any other nation and with iron ores regarded as inexhaustible, and many experts now declare that the end of both iron and coal is in sight.

The mere increase in our consumption of coal during 1907 over 1906 exceeded the total consumption in 1876, the Centennial year. The enormous stores of mineral oil and gas are largely gone. Our natural waterways are not gone, but they have been so injured by neglect, and by the division of responsibility and utter lack of system in dealing with them, that there is less navigation on them now than there was fifty years ago. Finally, we began with soils of unexampled fertility and we have so impoverished them by injudicious use and by failing to check erosion that their crop producing power is diminishing instead of increasing. In a word, we have thoughtlessly, and to a large degree unnecessarily, diminished the resources upon which not only our prosperity but the prosperity of our children must always depend.

We have become great because of the lavish use of our resources and we have just reason to be proud of our growth. But the time has come to inquire seriously what will happen when our forests are gone, when the coal, the iron, the oil, and the gas are exhausted, when the soils shall have been still further impoverished and washed into the streams, polluting the rivers, denuding the fields, and obstructing navigation. These questions do not relate only to the next century or to the next generation. It is time for us now as a Nation to exercise the same reasonable foresight in dealing with our great natural resources that would be shown by any prudent man in conserving and widely using the property which contains the assurance of well-being for himself and his children.

The natural resources I have enumerated can be divided into two sharply distinguished classes accordingly as they are or are not capable of renewal. Mines if used must necessarily be exhausted. The minerals do not and can not renew themselves. Therefore in dealing with the coal, the oil, the gas, the iron, the metals generally, all that we can do is to try to see that they are wisely used. The exhaustion is certain to come in time.

The second class of resources consists of those which can not only be used in such manner as to leave them undiminished for our children, but can actually be improved by wise use. The soil, the forests, the waterways come in this category. In dealing with mineral resources, man is able to improve on nature only by putting the resources to a beneficial use which in the end exhausts them; but in dealing with the soil and its products man can improve on nature by compelling the resources to renew and even reconstruct themselves in such manner as to serve increasingly beneficial uses—while the living waters can be so controlled as to multiply their benefits.

Neither the primitive man nor the pioneer was aware of any duty to posterity in dealing with the renewable resources. When the American settler felled the forests, he felt that there was plenty of forest left for the sons who came after him.

When he exhausted the soil of his farm he felt that his son could go West and take up another. So it was with his immediate successors. When the soil-wash from the farmer's fields choked the neighboring river he thought only of using the railway rather than boats for moving his produce and supplies.

Now all this is changed. On the average the son of the farmer of to-day must make his living on his father's farm. There is no difficulty in doing this if the father will exercise wisdom. No wise use of a farm exhausts its fertility. So with the forests. We are over the verge of a timber famine in this country, and it is unpardonable for the Nation or the States to permit any further cutting of our timber save in accordance with a system which will provide that the next generation shall see the timber increased instead of diminished. Moreover, we can add enormous tracts of the most valuable possible agricultural land to the national domain by irrigation in the arid and semiarid regions and by drainage of great tracts of swamp land in the humid regions. We can enormously increase our transportation facilities by the canalization of our rivers so as to complete a great system of waterways on the Pacific, Atlantic, and Gulf coasts and in the Mississippi Valley, from the Great Plains to the Alleghenies and from the northern lakes to the mouth of the mighty Father of Waters. But all these various uses of our natural resources are so closely connected that they should be coordinated, and should be treated as part of one coherent plan and not in haphazard and piecemeal fashion.

It is largely because of this that I appointed the Waterways Commission last year and that I have sought to perpetuate its work. I wish to take this opportunity to express in heartiest fashion my acknowledgment to all the members of the Commission. At great personal sacrifice of time and effort they have rendered a service to the public for which we can not be too grateful. Especial credit is due to the initiative, the energy, the devotion to duty and the farsightedness of Gifford Pinchot, to whom we owe so much of the progress we have already made in handling this matter of the coordination and conservation of natural resources. If it had not been for him this convention neither would nor could have been called.

We are coming to recognize as never before the right of the Nation to guard its own future in the essential matter of natural resources. In the past we have admitted the right of the individual to injure the future of the Republic for his own present profit. The time has come for a change. As a people we have the right and the duty, second to none other but the right and duty of obeying the moral law, of requiring and doing justice, to protect ourselves and our children against the wasteful development of our natural resources, whether that waste is caused by the actual destruction of such resources or by making them impossible of development hereafter.

Any right thinking father earnestly desires and strives to leave his son both an untarnished name and a reasonable equipment for the struggle of life. So this Nation as a whole should earnestly desire and strive to leave to the next generation the national honor unstained and the national resources unexhausted. There are signs that both the Nation and the

States are waking to a realization of this great truth. On March 10, 1908, the supreme court of Maine rendered an exceedingly important judicial decision. This opinion was rendered in response to questions as to the right of the legislature to restrict the cutting of trees on private land for the prevention of droughts and floods, the preservation of the natural water supply, and the prevention of the erosion of such lands, and the consequent filling up of rivers, ponds, and lakes. The forests and water power of Maine constitute the larger part of her wealth and form the basis of her industrial life, and the question submitted by the Maine senate to the supreme court and the answer of the supreme court alike bear testimony to the wisdom of the people of Maine, and clearly define a policy of conservation of natural resources, the adoption of which is of vital importance not merely to Maine but to the whole country.

Such a policy will preserve soil, forests, water power as a heritage for the children and the children's children of the men and women of this generation; for any enactment that provides for the wise utilization of the forests, whether in public or private ownership, and for the conservation of the water resources of the country, must necessarily be legislation that will promote both private and public welfare; for flood prevention, water power development, preservation of the soil, and improvement of navigable rivers are all promoted by such a policy of forest conservation.

The opinion of the Maine supreme bench sets forth unequivocally the principle that the property rights of the individual are subordinate to the rights of the community, and especially that the waste of wild timber land derived originally from the State, involving as it would the impoverishment of the State and its people and thereby defeating one great purpose of government, may properly be prevented by State restrictions.

The court says that there are two reasons why the right of the public to control and limit the use of private property is peculiarly applicable to property in land: "First, such property is not the result of productive labor, but is derived solely from the State itself, the original owner; second, the amount of land being incapable of increase, if the owners of large tracts can waste them at will without State restriction, the State and its people may be helplessly impoverished and one great purpose of government defeated. * * *

We do not think the proposed legislation would operate to 'take' private property within the inhibition of the Constitution. While it might restrict the owner of wild and uncultivated lands in his use of them, might delay his taking some of the product, might delay his anticipated profits and even thereby might cause him some loss of profit, it would nevertheless leave him his lands, their product and increase, untouched, and without diminution of title, estate, or quantity. He would still have large measure of control and large opportunity to realize values. He might suffer delay but not deprivation. * * * The proposed legislation * * * would be within the legislative power and would not operate as a taking of private property for which compensation must be made."

The court of errors and appeals of New Jersey has adopted a similar view, which has recently been sustained by the Supreme Court of the United States. In delivering the opinion of the Court on April 6, 1908, Mr. Justice Holmes said: "The State as quasi-sovereign and representative of the interests of the public has a standing in court to protect the atmosphere, the water, and the forests within its territory, irrespective of the assent or dissent of the private owners of the land most immediately concerned. * * * It appears to us that few public interests are more obvious, indisputable and independent of particular theory than the interest of the public of a State to maintain the rivers that are wholly within it substantially undiminished, except by such drafts upon them as the guardian of the public welfare may permit for the purpose of turning them to a more perfect use. This public interest is omnipresent wherever there is a State, and grows more pressing as population grows. * * * We are of opinion, further, that the constitutional power of the State to insist that its natural advantages shall remain unimpaired by its citizens is not dependent upon any nice estimate of the extent of present use or speculation as to future needs. The legal conception of the necessary is apt to be confined to somewhat rudimentary wants, and there are benefits from a great river that might escape a lawyer's view. But the State is not required to submit even to an æsthetic analysis. Any analysis may be inadequate. It finds itself in possession of what all admit to be a great public good, and what it has it may keep and give no one a reason for its will."

These decisions reach the root of the idea of conservation of our resources in the interests of our people.

Finally, let us remember that the conservation of our natural resources, though the gravest problem of to-day, is yet but part of another and greater problem to which this Nation is not yet awake, but to which it will awake in time, and with which it must hereafter grapple if it is to live—

the problem of national efficiency, the patriotic duty of insuring the safety and continuance of the Nation. When the people of the United States consciously undertake to raise themselves as citizens, and the Nation and the States in their several spheres, to the highest pitch of excellence in private, State, and national life, and to do this because it is the first of all the duties of true patriotism, then and not till then the future of this Nation, in quality and in time, will be assured.

Printed in Dunstable, United Kingdom

91219478R00037